DUST COLLECTION

RECOMMENDATIONS FOR HOM

First published 1991 by
Woodstock International, Inc.
Reprinted 1993, 1995, 2002

© 1991 Woodstock International, Inc.
P.O. Box 2309, Bellingham, WA 98227

Printed in U.S.A.

FOURTH EDITION

WOODSTOCK INTERNATIONAL, INC.

Item Number W1050

PREFACE

A considerable amount of consumer interest regarding design, installation and operation of dust collection systems has brought about the need to discuss general design considerations and the use of dust collection accessories in small home-shop environments.

Designing a dust collection system for efficient dust removal can be very complicated and is dependent upon many variables. Our intention is not to bog you down in too many details regarding the calculations involved in sizing a dust collector for a small shop system. However, it is important to know the relationships between different variables in order to better judge the relative efficiency of one design over another.

If your needs require a complex, industrial system, we recommend that you contact a professional design service. The higher costs involved with industrial systems make a design service very cost effective. Industrial applications also involve many more variables and trade-off decisions as well as strict compliance with OSHA and other regulatory agencies. There are many engineering consulting companies available that specialize in air and dust handling system design.

This handbook does not attempt to describe every aspect of safety, implementation and operation of any particular home-shop dust collection system. Your particular system must adhere to all rules and regulations set by The National Fire Protection Agency (NFPA), National Electric Code (NEC), Occupational Safety & Health Association (OSHA) and any other federal, state or local governing codes and requirements where applicable.

IMPORTANT: The information contained in this handbook is a recommended procedural method for designing, installing and operating a simple home shop dust collection system and is offered as a guide only. Woodstock International, Inc. does not assume any liability regarding the interpretation of this information. You are individually responsible for the safety and design of your particular dust collection system.

Table of Contents

PAGE

GETTING STARTED

The benefits of dust collection in a home-shop environment are readily apparent. Uncontrolled dust poses a serious potential health risk, not to mention the nuisance of dust in the shop and home. A home shop dust collection system can be as simple as connecting a mobile dust collector to a single woodworking machine like the one pictured below, or as complex as ducting multiple woodworking machines to a stationary collector. In order to consider a dust collection system for your home shop, you should have an idea of what you would like to accomplish in terms of collecting and transporting wood dust and chips.

Obviously, simple systems require less expense and are generally easier to set up and operate than larger, more complicated systems. However, simple systems may not conveniently handle all of the dust your shop may produce. Larger systems, on the other hand, may certainly do the job, but the extra expense may not justify the extra capacity. Balancing cost against the dust collection requirements for your shop should be a primary consideration when designing and building a dust collection system.

The key to building a relatively inexpensive dust collection system is to design the most efficient duct system possible. However, when designing an efficient duct system, there are many interrelated variables that must be considered. This handbook attempts to define many of the variables associated with building an efficient dust collection system. These variables can then be weighed according to the importance of your particular preferences and shop layout.

Last but not least, this handbook aims to simplify the complex science of dust collection so that you can spend more time on the projects that actually create the dust. Good luck and enjoy!

SYSTEM COMPONENTS

All basic dust collection systems include a dust collector and some type of duct system. The dust collector produces airflow by creating negative air pressure with a motor driven impeller. Air and dust travel in the direction of negative air pressure. The dust is collected in a drum or a collection bag, and the air is exhausted through a filter bag system.

The duct system attaches to the dust collector and is the means of conveying air, dust and chips from the woodworking machines to the dust collector. The duct system includes various piping, fittings and hoods. Duct systems range in size and complexity from simple one-machine systems to very large systems serving multiple machines.

More complex duct systems usually have a trunk line or mainline with branch lines running from the mainline to each machine. The mainline and branch lines are generally fixed to the ceiling or walls of the shop and include various fittings that allow directional changes or provide control functions. Fittings that provide directional changes include elbows, T's and Y's; while fittings that provide control functions include reducers and blast gates. Blast gates control airflow at each machine as needed and reducers are used to control the rate of air movement or air velocity within a duct.

Ducts are secured to each woodworking machine by way of some type of hood. The purpose of hoods, including floor sweeps, is to capture and direct dust and chips into the system. Some woodworking machines have built-in hoods, while others do not. For machines without standard hoods, accessory hoods can be purchased or shop fabricated. To determine inlet diameters for accessory hoods, follow the procedure described in this handbook for calculating branch duct diameters for individual woodworking machines. For those machines equipped with factory hoods, the branch line diameter should match the hood outlet diameter.

OVERVIEW OF DUST COLLECTORS

There is a wide variety of dust collectors on the market today that are all targeted for the home shop. They range in size from shop type vacuums with augmented filter bags to small scale industrial systems with cyclone separators. Naturally, these dust collectors have a comparatively wide range of dust collecting capabilities. However, all dust collectors should be rated by volume of air movement in cubic-feet-per-minute (CFM) at a given static pressure (SP) for comparison. Static pressure, in simple terms, is a measure of resistance to airflow. If airflow resistance created by any duct or filter system is greater than the static pressure rating of the dust collector, the volume of airflow will be less than the CFM dust collector rating. On the other hand, if the actual duct system and filter static pressure value is less than the dust collector static pressure value, the dust collector will move all of its rated volume of air.

The necessary CFM of air needed for most systems is easily determined since each woodworking machine produces a maximum volume of chips or sawdust at a given rate. In this handbook, these sawdust volume rates have been converted to airflow volume rates and have been tabulated for each type of woodworking machine.

Static pressure loss, or resistance to airflow, is caused by friction against the walls of the ducts, turbulence due to changes in air direction caused by fittings, and by physical restrictions such as clogged filter bags. The static pressure rating for any dust collector is measured in inches of water gauge. **This does not mean that dust collectors are designed to move water.** This unit of measurement is only used as a standard to rate the relative power of each dust collector. Dust collectors with higher static pressure ratings generally have the ability, or power, to move their given volume of air through more elaborate or less efficient systems.

Since there are so many different types and sizes of dust collectors available, you should first consider your specific needs by planning your dust collection system on paper. You can then shop for the collector that will satisfy the needs of your particular application.

There are two basic types of dust collectors available for home shop applications. These are single-stage collectors like the one pictured below and two-stage collectors. Each type has advantages and disadvantages.

Single-stage dust collectors create negative air pressure with an in-line impeller. Dust and chips are exhausted into a collection and filter bag after passing through the impeller. Precautions should be taken so that objects that are too large to pass through the impeller are not collected into the system. Two-stage collectors also create negative air pressure with an impeller, but heavier chips and objects drop into a first-stage collection container before entering the impeller. The finer dust then passes through the impeller and is collected in a second-stage filter bag. Single-stage collectors are generally easier to empty since the collection bag can be readily removed. Cleaning a two-stage collector is usually more difficult since the drum must be emptied after the motor and impeller housing are removed. In any event, the style selected is usually a matter of personal preference.

In a small home shop, the dust collector motor should have a totally enclosed fan cooled (TEFC) enclosure to prevent airborne dust near the dust collector from circulating through the motor.

All self contained dust collectors must also be routinely emptied. Dust collectors with filter bags and dust storage drums become increasingly less efficient as dust and chips fill the containers. Filter bags also lose efficiency as they become clogged with fine dust and should be routinely cleaned or safely shaken outdoors. **Always wear a respirator when emptying or cleaning a dust collector.**

Finally, with all dust collectors, there is a risk of fire hazard if sparks from metal striking metal or from abnormal cutting friction are drawn into the dust collection system. Sparks, fanned by an abundance of oxygen and fueled by wood dust have the potential to smolder and ignite into flame. If you ever suspect that sparks were generated and were drawn into the dust collection system, shut down the system and empty the dust collector storage drum or bag into a safe, air-tight container.

OVERVIEW OF AIR FILTERS

Primarily used to augment the capabilities of a dust collector, overhead air filters (shown below) remove very fine dust particles that remain suspended in the air. Fine dust suspension is a result of incomplete capture at the dust source and ultra-fine dust passing through the dust collector filter bag. Air filters work by drawing in airborne dust particles and trapping them in a series of filters. The processed air is thus cleaned and forced back into the shop to circulate. As this filtering process works continuously, the air in the shop is changed many times per hour.

Overhead air filters are extremely important for respiratory safety. Although the woodworker may wear a dust mask and have a dust collector in place, fine dust will hang in the air for long periods of time after cutting operations. This air can be easily inhaled.

From another standpoint, overhead air filters are also valuable in that they circulate heated/air-conditioned shop air that would otherwise be ventilated with outside air. This is a big benefit for any shop that expends a great amount of resources for temperature control.

An air filter should be selected based on the size of shop it needs to service. Use the formula below to determine the right air filter you need (rated in CFM) to reach the recommended air changes per hour.

Finally, the most important point to remember about overhead air filters is that they are a supplemental device. Air cleaners should be used with, but should never take the place of, a dust collector and a dust mask.

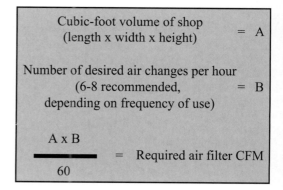

Cubic-foot volume of shop
(length x width x height) = A

Number of desired air changes per hour
(6-8 recommended, = B
depending on frequency of use)

$$\frac{A \times B}{60} = \text{Required air filter CFM}$$

DUCT MATERIAL

In most commercial workshops, the mainline and branch lines are usually metal pipe. Flexible hose is then used to connect each machine to the branch lines. In the case of small home shops, flexible hose may be used for both mainline and branch line ducts. Plastic piping is also a popular duct material for home shops, and as we will point out, each type has advantages and disadvantages.

METAL PIPE

There are many kinds of metal pipe available such as stove pipe, heating, ventilating and air-conditioning (HVAC) pipe, and pipe designed specifically for dust collection. Advantages of metal pipe include the fact that it is a conductor and does not contribute to static electrical charge build-up. However, static charges are still produced when dust particles strike other dust particles as they move through the duct. Since metal pipe is a conductor, it can be grounded quite easily to dissipate any static electrical charges.

One consideration, however, is that metal pipe is generally more expensive than plastic pipe, and it is usually not airtight unless specifically manufactured for dust collection. Specially manufactured metal pipe, on the other hand, is quite expensive. Metal pipe is also generally more difficult to cut and assemble. If you end up using metal pipe, make sure that it is 26 gauge or heavier. Using lighter gauge metal pipe poses the risk of collapsing and ruining your system.

FLEXIBLE HOSE

Flexible hose is available in many styles and sizes. Flexible rubber hose, poly-ethylene, plastic flex-hose and other flexible ribbed hose is generally used for short runs, small shops and at rigid duct-to-tool connections. There are many different types of flex-hose on the market today. These are manufactured from material such as polyethylene, PVC, cloth hose dipped in rubber and even metal, including steel and aluminum.

There are also many kinds of pure plastic flexible hose, such as non-perforated drainage type hose and dryer vent hose. Drainage type hose, while being economical, does not quite have the flexibility desired for a dust collection system. The inside of the pipe is also deeply corrugated and can increase static pressure loss by as much as 50% over smooth-walled pipe. Dryer vent hose, while being completely flexible, is non-resistant to abrasion and has a tendency to collapse in a negative pressure system.

If using flex-hose, you should choose one of the many types that are designed specifically for the movement of solid materials, i.e., dust, grains, and plastics. However, the cost of specifically designed flexible piping can vary greatly. Polyethylene hose is well suited for the removal of particulate matter, especially sawdust, since it is durable, relatively smooth on the inside and completely flexible. Polyethylene is also very economical and available in a wide variety of diameters and lengths for most applications.

Non-conducting flexible hose, no matter what type (including wire reinforced), must be completely grounded against static electrical charge build-up. A further discussion of grounding a duct system is presented in a later section of this handbook.

PLASTIC PIPE

Since plastic pipe is so common in agriculture, construction and general industry for conveying solids and liquids, it is only natural that it may also be used to handle wood dust. The popularity of plastic pipe is due to the fact that it is an economical and readily available product. Plastic pipe is also simple to assemble and is easily sealed against air loss.

The primary disadvantage of plastic pipe for dust collection, whether black ABS or white PVC (and even rubber), is the inherent danger of static electrical charge build-up. Since plastic is an insulator, static electricity is generated as dust particles flow against the walls of the pipe. It is very important when using insulating-type materials in a dust collection system, that no matter what the type, they be grounded to dissipate static charge build-up.

The secondary disadvantage of plastic pipe is its adaptability. In most cases, plastic pipe will not mate directly with common metal or plastic dust collection fittings or with flexible hose. A common method to overcome this situation is to wrap duct tape around the outside of the dust fitting to increase its diameter, then friction fit the fitting inside of the plastic pipe. Other methods call for making donuts or other custom adapters from wood, then securing the fittings with screws. These methods may not be pretty, but they usually work.

COMMON FITTINGS AND COMPONENTS

The most common blast gates, dust hoods, adapters and other fittings are made of ABS plastic for toughness and high wear resistance. Although they can be used with a variety of different piping systems, they are designed to be used in conjunction with flexible hose. Unlike standard plumbing PVC or ABS fittings which fit over solid pipe, these fittings are made to fit inside flexible hose and are securely clamped in place with adjustable hose clamps. Clamping with hose clamps eliminates the need for protruding screws and unreliable tape to connect fittings to piping.

Friction fit 2½" to 3" adapters are designed to fit most home shop vacuum hose and attachments. These adapters will allow dry dust pick-up with a standard duct system while utilizing special attachments such as nozzles and floor sweepers. It is important to note that the flex-hose branch duct that is connected to any attachment must be properly grounded against static electricity.

Adapters that connect special fittings to rigid plumbing pipe or machine hood diameters larger than 4" can be fabricated in the home shop out of wood. These wooden adapters are shaped like donuts (see Figure 1). The outside diameter equals the inside diameter of the larger pipe or hood. The inside diameter of the adapter equals the outside diameter of the smaller pipe or hood. These wooden adapters can be secured to the duct material or fittings with screws.

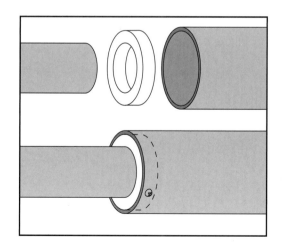

FIGURE 1

ILLUSTRATES A WOODEN DONUT USED
TO ADAPT DIFFERENT SIZED PIPING AND
FITTINGS.

Other fittings such as 6" to 4" reducers are commonly used in the HVAC industry and are available at heating and cooling supply centers. These supply centers are a good source for other off-the-shelf types of specialty fittings. Another good source, sheet metal fabricators can build most any type of fitting to your specifications. Many HVAC contracting companies have their own sheet metal fabricating shops, or they can refer you to a shop that will do custom work at reasonable rates. Refer to the Heating Contractors Section in your local Yellow Pages.

Clean-outs can be placed in strategic locations throughout the duct system. For example, a Y can be used at the end of the mainline instead of an elbow to the last branch. This not only provides a more gradual direction change, but the open end of the Y can be capped with a blast gate that can be opened for visual inspection and provide easy clean-out if necessary. **See Figure 2**. This Y will also be the most logical place to expand the duct system when the need arises, as long as the dust collector can handle the extra capacity.

FIGURE 2

Shows a Y at the end of a main-line fitted with a blast gate.

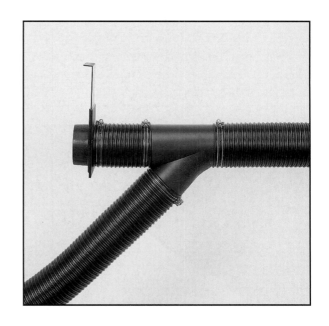

The following are descriptions and pictures of common dust collection fittings and components.

Y-Fittings: Y's provide a gradual inter-section of one duct into another duct.

T-Fittings: T's provide a 90° intersec-tion of one duct into another. T's should only be used when duct routing con-straints will not allow the use of Y's.

Elbows: Elbows allow the duct to change direction. For efficient airflow, elbows should be smooth walled with a minimum center-line radius of 1.5 times pipe diameter.

Splices: Splices allow easy connection of one duct to another when two duct sections must be joined together. They are used when making long runs or when utilizing short pieces of duct.

Reducers: Reducers change duct diameter. They are used to downsize a duct to increase the velocity necessary to carry dust in suspension.

Combination Adapters: Combination adapters allow specialty sized hoses and fittings such as home shop vacuum attachments to be connected to standard sized dust collection hose.

Universal Adapters: Universal adapters allow the user to cut away unneeded steps to create a custom combination adapter.

Floor Sweeps: Floor sweeps direct dust into the system when sweeping loose dust from the shop floor. They can be conveniently located and are controlled by blast gates.

Blast Gates: Blast gates provide airflow control for each machine in a system. By opening the blast gate for the machine to be used, and closing those for machines not in use, airflow will be directed to the dust source.

Universal Dust Port: Can be mounted to many different types of dust-producing machines. The compact mounting flange and angled port can be adapted to bandsaws, stationary sanders, router tables or shapers. The 2½" port is sized for home shop vacuums and can be adapted for central dust collection systems.

Hose Clamps: Hose clamps are used to secure flexible hose to various fittings and hoods while ensuring an air tight system. They are offered in many styles, the most popular being wire and band. Style choice is a matter of personal preference.

Quick Connect: Quick connects are tapered to allow easy connection to machine hoods or system components with a friction fit.

Hose Hangers: Hose hangers are a quick and easy means of mounting ductwork to walls or ceilings. Simply place over the duct and screw the hangers to joists or studs through the provided holes.

Jointer Dust Hood   Universal Dust Hood   Table Saw Dust Hood

Hoods: Hoods capture dust at the source and provide easy connection to the duct system. The jointer hood adapts to larger openings in cabinet type jointers. The table saw hood on the right attaches to open frame contractor type table saws. And, the universal type hood easily attaches to most cabinet type table saw stands.

> **⚠WARNING**
> When installing dust collection components such as hoods, they must not hinder or impede machine or safety guard operation.

Dust Collection Separators: Dust Collection Separators increase holding volume and filtering capacity of standard single-stage systems by removing larger chips and wood particles from the airflow before they reach the dust collector. The Dust Collection Separator attaches to a 30-gallon metal garbage can and is connected to the mainline ducting near the dust collector. As negative pressure passes through the Dust Collection Separator, a cyclone action occurs and heavier particles are allowed to drop to the bottom of the garbage can. The airflow then carries the remaining dust and lighter particles to the collector's containment bags. In essence, the Dust Collection Separator turns the ordinary single-stage system into a much more effective and capacious two-stage system. The separator also reduces the potential for fire hazard by removing metal objects such as nails from the system before they hit he impeller. Saw dust and chips are easily removed by lifting the separator and emptying the metal can. We highly recommend using a dust collection separator.

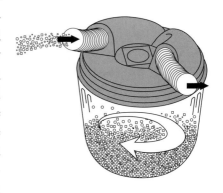

Remote Control Power Switches:
Remote control power switches are an easy way to turn on your dust collector from anywhere in the shop. A remote control saves the user from physically walking over and flipping the dust collector switch every time he or she starts a machine.

Micro-Porous/Oversized Filter Bags:
Micro-porous/oversized filter bags offer increased airflow, decreased differential pressure and improved overall air cleaning efficiency. Common sizes include 5, 1 and .3 micron-rated bags. The lower numbers represent better filtering capabilities. Often micro-porous bags are oversized to increase the volume of air to make up for the smaller pores in the surface. Because the most harmful dust particles are under 10 microns, these bags can be an important health consideration, especially if the dust collector is not used in conjunction with an air filter.

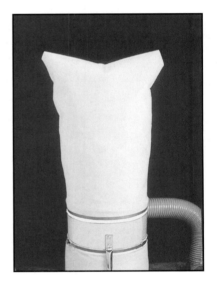

SYSTEM DESIGN

The first and most important step in designing a dust collection system is to plan ahead. The best way to plan is to draw a simple bird's-eye view of your shop and sketch in the following:

1. Your desired location for the dust collector.
2. The location of each woodworking machine in the system. Woodworking machines that produce the most chips and sawdust—such as planers, shapers and bandsaws—should be located nearest to the dust collector.
3. The location of the mainline duct and each branch line. Your system should be designed so that it has the shortest mainline run possible, with short secondary branch ducts.
4. The location of any obstructions (floor joists, beams, posts, fixtures, etc.) which will require special duct routing.

Your workshop sketch should be drawn to scale to aid in estimating the length of the mainline and branch line runs. Convenient scales to use are ¼":1', ⅜": 1' or ½":1'. The larger ½":1' scale will provide greater detail when sketching smaller shops. For convenience, scale planning grids are provided in the back of this handbook. They can be copied to produce a number of different plans or taped together to increase the overall plan size.

There are many ways to design a mainline and branch duct system. One design type, as illustrated in **Figure 3** on page 20, shows a mainline running down the center and along the ceiling of a hypothetical shop. This basic design type will also work well if conditions allow running the duct system under a shop floor such as in a crawl space or if planning and installing piping prior to pouring a concrete floor.

Advantages to having the duct system located under the floor include: 1) the duct system will not interfere with movement and operations within the shop, and 2) the duct system is located much closer to the dust-producing machines, which translates into greater dust collection efficiency.

A primary disadvantage to an under-floor duct system is less flexibility for expansion or changing machine placement within the shop, particularly if the duct system is set in concrete. Provisions must also be made in the event of duct system clogging in inaccessible areas. This could be as simple as installing clean-outs in strategic locations. Clean-outs are just Y's or T's with the lateral port extended so it is accessible and capped off with a blast gate. If a clog occurs, the nearest clean-out can be opened and the clogged material can then be manually dislodged.

Whether you choose to install your duct system overhead or under the floor, the mainline will run across the shop, and each woodworking machine will be accessed from either side of the mainline by branch ducts. Mainlines and branch ducts should be straight, and branch ducts should intersect with the mainline at 45° for 45° Y connections and 90° for T connections. Once again, Y connections are more efficient than abrupt T connections, but may not always be feasible depending upon your application.

Another variation to the centrally-located mainline theme puts the dust collector in the corner of the shop, while the mainline runs along the ceiling next to the walls or runs diagonally across the center of the shop. **Figure 4** on page 21 shows the mainline located around the inside perimeter of our hypothetical shop. Since woodworking machines are often located against shop walls, running the mainline along the perimeter may be more convenient. This design usually requires a longer mainline run, but may result in shorter branch lines. This will also allow the dust collector to be located in a corner of the shop. Another advantage to this design allows the mainline duct to be located at machine height instead of overhead. However, these variations may or may not be as efficient as centering the mainline in the shop, depending on your shop size and layout.

The relative efficiency of any system is dependent upon the limiting or least efficient branch run. The efficiency of each branch run is dependent upon its length (by diameter class) from the dust collector to the dust producing machine and the number, type and size of fittings in the branch run. When comparing different duct system designs, consider the limiting branch duct efficiency and personal preference.

When you begin planning, we recommend that you draw your shop so each machine is located where you want it based upon preference and material processing efficiency and then experiment with the dust collector location and duct layout. You may need to produce a number of drawings to quantify and compare efficiency based upon scaled duct lengths, duct diameters and number and type of fittings.

For illustration purposes, our **Figure 3** sample sketch shows the mainline running horizontally along the ceiling. Branch ducts Y along the ceiling and drop down to each machine and floor sweep. A dust separator is used to collect the large wood chips and protect the dust collector impeller from any small nails that may get swept into the floor sweep. This is also an important safety precaution.

Horizontal duct lengths can be measured from the scale drawing and the length of the drops can be measured in the shop from each machine hood location, up to the ceiling. In our illustration, we separated length of drop measurements with an asterisk. At this time, do not worry about duct diameters. These will be determined later.

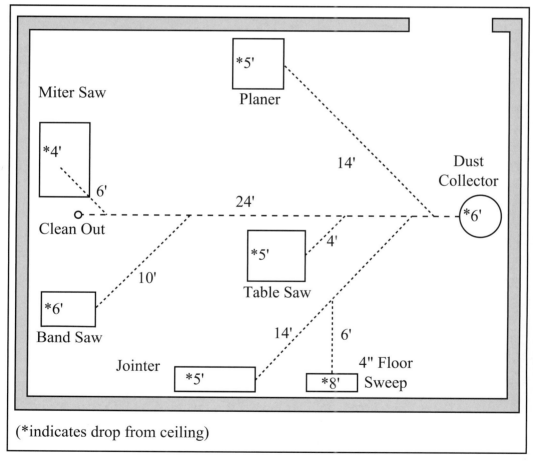

FIGURE 3.

ILLUSTRATES A SIMPLE BIRD'S-EYE VIEW OF A 30' X 24' SHOP WITH
THE MAINLINE DUCT RUNNING DOWN THE CENTER OF THE SHOP.

Our **Figure 4** sample sketch shows the mainline running horizontally along the walls. Branch ducts T or Y from the mainline to each machine. Again, a dust separator is used to collect the large wood chips and protect the dust collector impeller from any small nails that may get swept into the floor sweep.

As with **Figure 3**, horizontal ducts can be measured from the drawing and branches can be measured to each machine hood. This layout has one advantage in that the duct work stays closer to the machines, thus, making it more efficient.

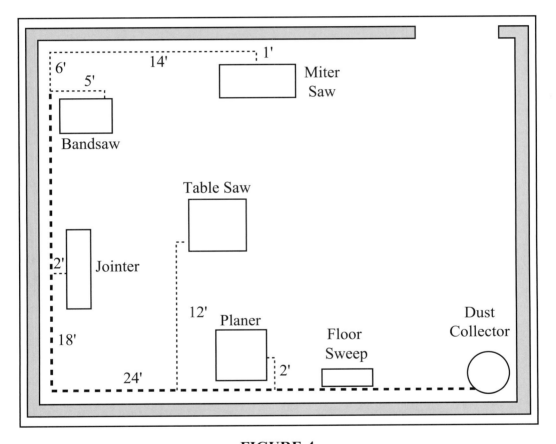

FIGURE 4.
ILLUSTRATES A SIMPLE BIRD'S-EYE VIEW OF A 30' x 24' SHOP WITH
THE MAINLINE DUCT RUNNING AROUND THE INSIDE PERIMETER OF THE SHOP.

Determining a location for the dust collector is an important part of any system design. A popular location for the dust collector is in a separate, but adjacent area to free up work space, reduce noise in the shop and eliminate very fine dust exhausted through the filter bag. On/off switching for a remote dust collector is relatively simple and inexpensive, depending upon the degree of sophistication. Contact a licensed electrician for more information about wiring a remote switching setup. On the other hand, remote controlled power switches are an excellent alternative because they are economical and easy to setup by yourself.

If you are planning to locate your dust collector outside of the shop area, please consider the following:

1. Do not locate a dust collector in a room that contains gas appliances with pilot lights or open flames. There is a risk of explosion due to dust dispersal into the air from: 1) very fine dust escaping through the filter bags during normal operation, 2) the accidental chance of a collection or filter bag becoming loose during operation, or 3) loose dust escaping during normal cleaning procedures.

2. If locating the dust collector in a separate, enclosed room, precautions should be taken to minimize the risk (however remote) of a dust explosion. In order for a dust explosion to occur, there must be a sufficient amount of fine dust in suspension and an ignition source. Eliminating any of these conditions will eliminate the risk of an explosion. To eliminate an ignition source, convert the dust collector motor to an explosion proof motor and ensure that static electricity is eliminated or adequately grounded. An explosion proof motor eliminates the risk of a dust explosion caused by the electrical spark necessary to energize the motor starter winding. Reducing the amount of fine dust in the air is another option. You may consider circulating the air in the room with clean filtered air during dust collector operation.

3. If an equal amount of air removed by the dust collector is not returned back to the room by way of either filtered or fresh air, there will be a dramatic pressure difference between the room supplying the intake air and the point where air is exhausted. This pressure difference will contribute to a loss of dust collector efficiency, not to mention heat or air conditioning loss from within the shop. Always provide enough access for the same amount of unrestricted air to return back into the shop as there is leaving through the dust collection system. Air that is returned into the shop from an adjacent room containing the dust collector can be filtered by standard forced air furnace filters. Filtered return air will protect the shop environment from very fine dust escaping from the filter bag during operation. A filter frame or frames can be located in the common wall between the shop and dust collector. Each frame should be designed so the filter can be removed for cleaning and replacement. Furnace filters are available in a variety of sizes and work particularly well for these types of frames.

4. Any dust collector motor should be protected against the hazard of overheating due to starting failure or overloading. This protection may be a separate overcurrent device such as a motor starter complying with Article 430 of the National Electric Code, or the motor must have an integral manual reset.

5. Consider the dust separator as an important safety precaution against fire. If something metal (a nail) gets sucked into the system and hits the dust collector impeller, sparks may result. Even if these sparks don't cause a dust explosion, they could land in the collected dust and smolder for days before causing a fire. (This hazard is especially compounded when using a floor sweep.) A good dust separator will effectively trap the metal object before it reaches the dust collector impeller.

DUST COLLECTION NEEDS

Once you have a basic idea about where you want to place the dust collector in relation to the other machines in your shop, you should focus on determining duct diameters and fittings.

DETERMINING AIR MOVEMENT

Air movement is the first factor you need to know when calculating duct size. As previously stated, each woodworking machine produces a certain volume of sawdust and requires a minimum airflow in cubic-feet-per-minute (CFM) to move that sawdust. The required air movement necessary for individual woodworking machines can be obtained from the table below. Please note that the stated required airflow represents volumes for average home shop machinery.

Machine	Required Airflow (CFM)
Table Saw	350
Miter/Radial-Arm Saw	350
Jointer (6" and smaller)	350
Jointer (8"-12")	450
Thickness Planer (13" and smaller)	400
Thickness Planer (14"-20")	785
Shaper	350
Router (mounted to table)	200
Bandsaw	350
Lathe	350
Disc Sander (12" and smaller)	350
Disc Sander (13-18")	450
Belt Sander (6" and smaller)	450
Belt Sander (7"-9")	550
Drum Sander (200 sq. in. and smaller)	350
Drum Saner (201-400 sq. in.)	550
Floor Sweep 4"	350

TABLE 1.

LISTS MINIMUM CFM AIR MOVEMENT FOR TYPICAL MACHINE USE.

Make copies of your shop layout sketch for various types of labeling (or lightly pencil in the measurements). To help demonstrate each step, we will use our previous shop sketch* from **Figure 3**.

First, use the numbers on **Table 1** from the previous page to label each machine with its respective CFM requirement as shown in **Figure 5**.

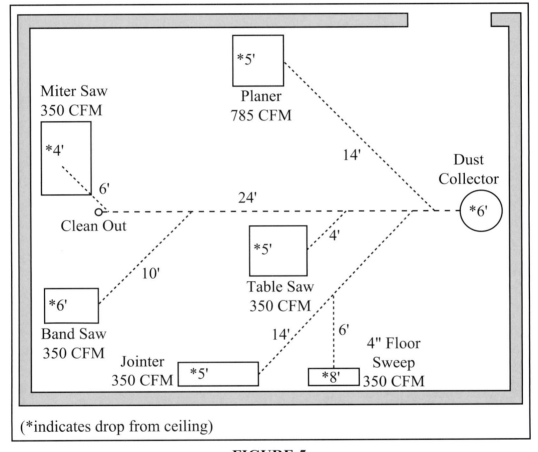

FIGURE 5.
Shows sketch with required CFM labeled for each machine.

* *Please keep in mind that our hypothetical shop sketch is for demonstration purposes only and is in no way indicative of how your final results should be. We chose this shop layout because it is large enough to demonstrate a multitude of options when designing a dust collection system.*

DETERMINING DUCT DIAMETERS

Once the minimum volume of air movement required for each machine has been determined, you must then determine duct size. This is critical to the performance of your dust collection system.

As a basic rule, **air velocity must not drop below 3,500 feet-per-minute (FPM) in the mainline and 4,000 FPM in branch lines** or wood chips and dust particles will begin to settle out of the air stream and collect in the bottom of the duct.

The most common error when designing a dust collection system is initially reducing the size of the duct system at the dust collector, thinking that an increase in air velocity is needed to move more dust particles. While velocity is certainly increased, wall friction is also increased dramatically and the driving force is less efficient. This does not mean that minimum air velocity is not important. Air velocity must be sufficient to maintain dust and chips in suspension.

Generally, the mainline should be as short as possible with a diameter as large as the dust collecting unit will allow. Air will move at a slower rate, but there will also be less air resistance.

The size of the mainline duct is determined by the maximum CFM air movement and the minimum required velocity. For example, if you plan to operate one machine at a time, the size of the mainline duct diameter will be dependent upon the machine with the greatest airflow requirement while maintaining minimum velocity. If you intend to operate two or more machines simultaneously, then the mainline must be sized for the combined CFM of air while maintaining minimum velocity. However, if a mainline is sized to handle a certain volume of air from two or more branches and the total volume is restricted by closing one or more branches, the volume of air entering the mainline may be insufficient to achieve the minimum mainline velocity.

Use **Table 2** to determine the appropriate duct size for the required machine CFM. As a general rule, always go down to the nearest whole number. (For example, if the table says to use 4.28 inch duct, use a duct diameter of 4 inches.)

Machine CFM Requirement	Main Line 3500 FPM Duct Dia. (inch)	Branch Line 4000 FPM Duct Dia. (inch)
300	3.97	3.71
350	4.28	4.01
400	4.58	4.28
450	4.86	4.54
500	5.12	4.79
550	5.37	5.02
600	5.61	5.25
650	5.84	5.46
700	6.06	5.67
750	6.27	5.86
800	6.48	6.06
850	6.67	6.24
900	6.87	6.42
950	7.06	6.60
1000	7.24	6.77
1200	7.93	7.42
1400	8.57	8.01
1600	9.16	8.57

TABLE 2.

LISTS MAIN AND BRANCH LINE SIZES FOR REQUIRED MACHINE CFM.

First determine the branch line diameter coming from each machine. If a machine comes with a dust port, use that size for the branch line or use an adapter.

Appendix 2 in the back of this handbook contains a helpful chart that shows how changes in duct diameter can change the potential airflow velocity. This chart may be useful when designing duct sizes for your system.

In our hypothetical shop in **Figure 6**, our ideal duct size for the planer is a 6" mainline with a 5" branch line. Our calculations show that the other machines would need a 4" mainline and branch line. Because there is the need for 6" duct in part of the main line, we have a problem. The 350 CFM will fall below the 3500 FPM of required velocity when it enters the 6" mainline. Our solution—if we open two 4" branches during operation, the 350 CFM will rise to 700 CFM. If we check 700 CFM on **Table 2** for a mainline diameter. The duct diameter is listed at 6.06", which would be rounded to 6". This works perfect!

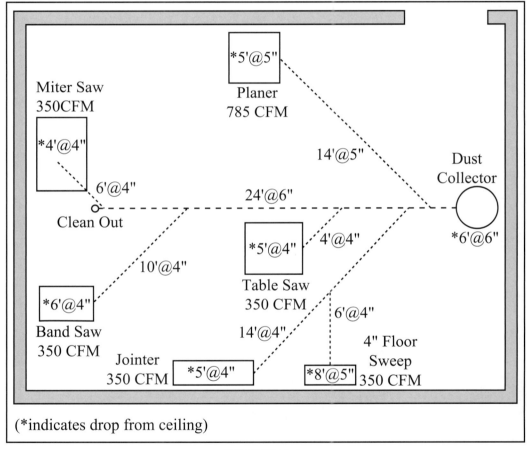

FIGURE 6.

BASIC DUCT SIZE LABELED FOR THE MAINLINE AND BRANCH LINES.

Now we have a good idea of our total length and diameter of duct work. The next step is to determine how to place the fittings and blast gates to make the dust collection system as efficient as possible.

PLANNING LAYOUT OF FITTINGS

Use your shop sketch (or one of the copies) to layout the fittings on the plan. **Figure 7** illustrates some of the ways fittings can be used. Below are some quick tips to keep in mind when designing your system:

- Directional changes should be kept to a minimum. The more elbows, T's and Y's there are from the dust collector to the end of a branch, the greater the static pressure loss.
- Try to use Y's rather than T's whenever possible.
- Gradual directional changes are more efficient than sharp turns.
- If you plan on using mostly rigid duct, place a small amount of flexible hose somewhere between the machine and the mainline duct—this will protect the rigid duct if the machine moves a little.
- A rigid 90° elbow is more efficient for making 90° turns than flexible pipe.
- Each branch should have a blast gate for maximum airflow control.
- The simpler the system, the more efficient and less costly it will be.
- Place the branch lines with the largest duct diameters closest to the dust collector.

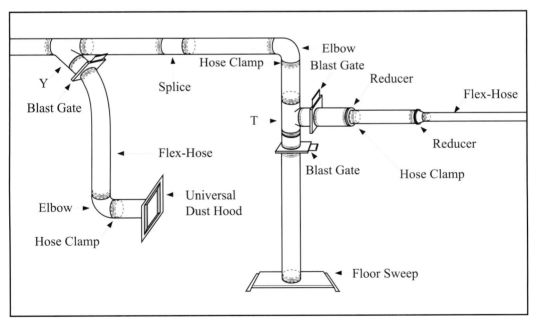

FIGURE 7.

Illustrates a typical dust collection set-up.

DETERMINING STATIC PRESSURE LOSS

The next step is to determine the maximum static pressure loss for your proposed system. To do this, you must calculate the least efficient run in your system—i.e., the run from the machine to the dust collector that has the longest combination of duct line and the largest amount of fittings. Use **Tables 3-5** to determine the static pressure loss for each component in your system.

For our hypothetical shop sketch, we mapped out the duct work and fittings to the miter saw, because it is the farthest away and most likely to have the maximum static pressure loss. **Figure 8** demonstrates this layout.

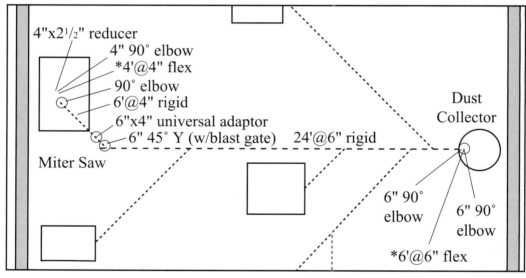

FIGURE 8.

LONGEST STRETCH OF FITTINGS, MAINLINE AND BRANCH LINE MAPPED
OUT FOR CALCULATING MAXIMUM STATIC PRESSURE LOSS FOR THIS SYSTEM.

Duct Diameter	Static Pressure Loss Per Foot of Rigid Pipe		Static Pressure Loss Per Foot of Flex Pipe	
	Main Lines at 3500 FPM	Branch Lines at 4000 FPM	Main Lines at 3500 FPM	Branch Lines at 4000 FPM
1"	.126	.170	.485	.625
2"	.091	.122	.35	.453
2.5"	.08	.107	.306	.397
3"	.071	.094	.271	.352
4"	.057	.075	.215	.28
5"	.046	.059	.172	.225
6"	.037	.047	.136	.18
7"	.029	.036	.106	.141
8"	.023	.027	.08	.108
9"	.017	.019	.057	.079

Table 3.

DUCT DIAMETER STATIC PRESSURE LOSS (IN INCHES) PER FOOT.

Fitting Diameter	90° Elbow	45° Elbow	45° Y-Fitting	30° Y-Fitting
3"	.47	.235	.282	.188
4"	.45	.225	.375	.225
5"	.531	.266	.354	.236
6"	.564	.282	.329	.235
7"	.468	.234	.324	.216
8"	.405	.203	.297	.189

Table 4.

FITTING STATIC PRESSURE LOSS (IN INCHES) PER FITTING.

Additional Factors	Approx. Static Pressure Loss
Dust collection separators (see page 16)	2"
Seasoned (used) dust collection filter	1"
Entry loss at large machine hood	2"

Table 5.

ADDITIONAL STATIC PRESSURE LOSS (IN INCHES) FACTORS.

The next step is to make a list of all the components mapped out for your least efficient run. Across from each component, make a space for the static pressure loss. The list for our hypothetical shop went like this:

4" 90° elbow	.45
4'@4" flex (branch line)	1.12
4" 90° elbow	.45
6'@4" rigid (branch line)	.45
6" 45° Y	.329
24'@6" rigid (main line)	.888
6" 90° elbow	.564
6'@6" flex (main line)	.816
6" 90° elbow	.564
Static Pressure Loss For Miter Saw Line	5.631
Additional loss from seasoned filter	1.0

TOTAL STATIC PRESSURE LOSS FOR THIS LINE   6.631*

If you are unsure about which run is the least efficient, you can map out every duct run for every machine in this manner—just be sure to include everything from the machine to the dust collector. Compare the total number of each run to verify which one has the largest static pressure loss in your system. **The run with the highest number is your maximum static pressure loss.** When you have your maximum static pressure loss number, then you are ready to determine which dust collector is right for your situation.

The numbers in Tables 3-5 are displayed only as a general guide. Calculating exact static pressure of any system requires expensive equipment and complex mathematical formulas.

SIZING A DUST COLLECTOR

To determine the right size dust collector for your system, look at the manufacturer's ratings for **Air Suction Capacity** and **Static Pressure**.

The **Air Suction Capacity** for the dust collector must be higher than the machine that requires the highest CFM (or combination of machines if you're running more than one at a time). In the case of our hypothetical shop, the largest requirement is the planer at 785 CFM.

The **Static Pressure** rating for the dust collector must be higher than the maximum static pressure loss for your system. For our hypothetical shop, this number is 6.631".

Make sure that any dust collector you choose is rated higher on **BOTH** of these numbers. If possible, leave extra room for any unforeseen factors.

If you already own a dust collector, the required CFM of air movement and static pressure loss for your system should not exceed the air suction capacity and static pressure rating for that collector.

For our hypothetical shop, we looked at a 2 HP dust collector with a 1550 CFM air suction capacity and a 12.3" static pressure rating. The extra CFM air suction capacity is more than enough for our system, and the extra static pressure ability would leave over 5.6" for any unforeseen factors or the addition of a dust separator.

BUILDING YOUR SYSTEM

Careful planning prior to actual construction will make layout and installation much easier. The following information presents some useful ideas for building your dust collection system.

MAKING A MATERIALS LIST

Refer back to your scaled shop drawing for the layout plan. Start at the dust collector location and layout your runs with a chalk line or an extended pencil line along a straight edge. Mark your layout lines on the walls or ceiling. All ducts should run straight and level. Take measurements for your ducting off of your layout lines and deduct for any fittings. When deducting for fittings, do not subtract for the flanges since the duct material will either slide over or into the fitting, depending upon the type used. The following information presents some useful ideas for building your dust collection system.

To calculate the material needed to build your system, make a materials list. Refer back to your shop sketch and make a list of all fittings including Y's, T's, elbows, reducers, splices and blast gates. Floor sweeps are also very handy and can be added with a blast gate, provided the floor sweep duct doesn't exceed the capacity of the dust collector. Also, count the number of hose clamps by size, remembering to include one clamp per fitting opening. When figuring for flex-hose, make an allowance for extra flex-hose if machines will be mobile. You should also account for any additional material needed to route the duct behind or around existing shop fixtures such as heating vents, posts, beams and pipes, if necessary. Plastic pipe and flex-hose should not be located near, or in contact with, any heat source such as heating vents or hot water pipes.

INSTALLING THE DUCT SYSTEM

There are many ways to position and support a duct system along a wall or ceiling. The quickest method is to use hose hangers (**see page 15**). Another method that works well for all types of duct material is to use wood lath strips fastened to the ceiling or wall and positioned on your layout lines. **See Figure 9**. The lath can be any convenient thickness and width such as ¾" x 1½". The duct can be secured to the lath with baling type wire or self-locking plastic straps commonly know as cable ties. Cable ties are available at larger hardware stores, electronics stores and heating and cooling supply stores. If using baling wire, cut off usable lengths, wrap it around the duct and lath strip and twist the ends together with pliers. Cut off the excess and bend the twisted end flat against the duct, or position it so the sharp end does not pose a safety hazard. Provide enough baling wire or cable ties along the duct so it is adequately supported.

FIGURE 9.

SHOWS INSTALLING A WOOD LATH TO A CEILING JOIST.

This duct support system also works very well against smooth walls and ceilings since lath can be attached to ceiling joists or wall studs through the drywall. If it is necessary to run the lath parallel and between two wall studs or ceiling joists, the lath can be attached directly to the drywall with any of the variety of drywall anchors available. If mounting flush to a wall or ceiling, the back of the lath can be cross-kerfed to accept the wire or strap.

If ceiling joists and wall studs are exposed, the lath system still simplifies duct installation and can be directly attached to the framing members. If the duct layout dictates that any duct must run parallel and between two wall studs or ceiling joists, 2x4 blocking can be installed between the studs or joists to support the lath. Determine the number of blocks needed to support your duct section. A spacing of four feet between blocks should be adequate. To install the blocking, simply measure the space between the studs or ceiling joists and cut the number of 2x4's needed. Make each 2x4 flush with the studs or ceiling joists and fasten with nails or screws. **Figure 25** on page 50 shows a vertical branch duct supported by a lath strip fastened to two 2x4 blocks that are secured between two wall studs.

Once you have installed the wood lath, you should then connect and hang one section of pipe or flex-hose at a time, adding fittings and internal ground-wire as required. **See Figures 10-12.** Also, please refer to the *Duct Grounding* section for a discussion regarding grounding for static charge.

Flex-hose is designed to slip over various fittings and can be secured with appropriately sized hose clamps. This system provides for easy assembly and disassembly if needed and ensures an air-tight connection. Rigid piping should also be constructed so it can be easily disassembled. This usually requires screw connections. If using screws, choose screws of minimum length so that airflow is not impeded and chips and sawdust will not wrap around them. Self tapping screws make installation quick and easy. Also, rigid pipe will usually slip over most dust collection fittings and can connect to flexible pipe with a splice.

When constructing the duct lines, you should ensure that the pipe/hose sections and fittings can be taken apart for inspection or removal of any obstruction. When working with rigid plastic pipe, we recommend against permanently gluing pipe and fittings together. If you do choose to glue pipe and fittings together and clogging occurs, or if you wish to modify your permanently secured duct system, rigid plastic pipe can be cut and repaired with repair couplings. Finally, ensure that blast gates are conveniently located in each branch duct and securely anchored so they can be pushed/pulled open or closed without deflecting the duct.

FIGURE 10.
SHOWS ATTACHING A SECTION OF HOSE TO A Y FITTING.

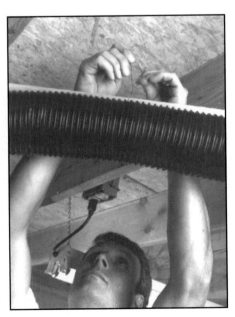

FIGURE 11.
SHOWS SECURING WIRE AROUND FLEX-HOSE AND WOOD LATH.
SELF-LOCKING PLASTIC TIES ARE ALSO VERY HANDY FOR THIS PURPOSE.

FIGURE 12.
SHOWS FASTENING FLEX-HOSE TO A BLAST GATE.

MACHINE HOODS

Dust hoods are the point of capture for sawdust entering the dust collection system. Some woodworking machines come standard with a machine hood, but many do not. If you add hoods, they should be mounted as close as possible to the point of dust generation for the greatest efficiency; however, normal machine operation and safety guard operation must not be affected by hood placement.

Installing hoods as close as possible to the source will also ensure that very fine dust, which is the primary contributor of respiratory problems, is captured. If very fine dust is not captured at the source in a closed shop environment, it will stay suspended in the shop air for a period of time, depending upon just how fine the dust is. Large dust particles will settle faster than fine dust particles. However, once fine dust is disturbed by normal shop operations, it will become airborne again, repeating the cycle. Of course, the greater the amount of suspended fine dust in the shop, the greater the health hazard. Naturally, capturing every bit of dust for every type of machine and every shop operation is impossible. Limiting the amount of dust that escapes from the dust collection system will limit the concentration and the health hazard.

Some woodworking machines may benefit from more than one hood, such as any machine producing a dust deflection and escape velocity greater than the dust hood and collection system can capture. A very good example of this is the table saw. Depending upon the blade position relative to the wood being cut, some dust is carried around and thrown up through the blade insert by the spinning blade. A secondary hood mounted above the blade in addition to the hood mounted below the blade will be more effective in capturing this dust. Stationary edge or belt sanders are another example of machines that could benefit from a secondary hood. Please note that the size of the branch duct for that machine should be sized to handle the two intake hoods. One blast gate can be used to control both hoods as long as the blast gate is located in the branch duct before the Y split.

Most suppliers offer a complete line of after-market hoods for common cabinet-style stand applications. These hoods can be mounted with screws or secured with double-sided carpet tape, which is available at most carpet supply stores. After-market hoods are easily mounted over existing dust ports as in the case of jointers, or if dust ports are not present, round or rectangular holes can be cut in the side of the machine cabinet. **See Figure 13**. To aid in cutting round or square holes in sheet metal cabinets, power sheet metal shears or nibblers can be rented at most rental centers to produce a neat, professional job. Bi-metal hole saws are also great for cutting round holes. The hole should be located as close to the bottom of the cabinet as possible and centered with the dust hood outlet. Minimum hole size should be no less than the hood outlet size.

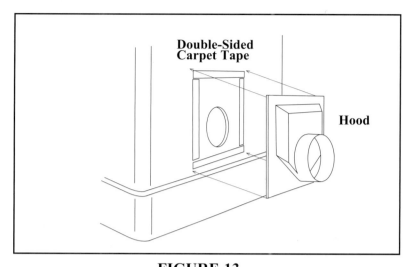

FIGURE 13.

ILLUSTRATES HOOD INSTALLATION USING DOUBLE SIDED CARPET TAPE.

If you do not want to permanently modify your cabinet style stand by cutting a hole in the sheet metal, after-market hoods can also be adapted to fit over existing openings such as dust clean-out access doors. An adapter is simply a shop-made plywood panel cut to fit the back of the hood. The hood can then be secured over the hole cut-out in the plywood panel. **See Figure 14.**

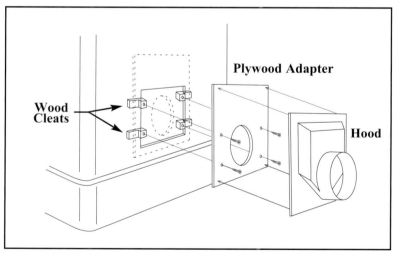

FIGURE 14.

SHOWS THE USE OF A PLYWOOD ADAPTER.

Most suppliers also offer hoods with larger flanges designed for use with open-frame contractor-type table saws. These hoods mount directly under the saw blade between the saw body and the stand. Once again, if the opening is too large for the hood to fit, a thin piece of plywood with a round or square hole cut-out can be mounted between the saw and stand. The hood can then be mounted over the hole in the plywood. **See Figure 15.**

Many cabinet style and contractor style table saws have openings in the stand or saw body to allow for motor movement when adjusting the angle of cut. These openings should be sealed as much as possible to limit the amount of escaping sawdust. Enclosures can be made out of sheet metal and designed so they do not impede motor movement. Enclosures should also be designed so they can be easily removed for inspection or maintenance. If the saw requires a box enclosure for the motor, as do most cabinet style stands, it too can be made by a sheet metal fabricating shop.

Sheet metal heating and cooling register boots can also be used or modified to work as hoods. These boots are available in a variety of sizes and shapes and fit 4" and 6" ducts. They are available at most hardware stores and heating and cooling supply stores.

As this section points out, jointers and table saws are fairly simple machines to adapt to a dust collection system, other machines like bandsaws, radial-arm saws, shapers and router tables may require a bit more creative ingenuity.

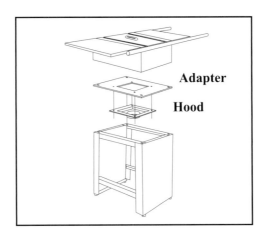

FIGURE 15.
SHOWS MOUNTING A HOOD FOR CONTRACTOR TYPE TABLE SAWS.

Radial-arm saws and shapers may require a hood that can be rotated in the direction of sawdust and chip deflection. These hoods could be as simple as a four-sided wooden box with an aftermarket hood attached to the back side. The open ended box is an extension of the hood and captures and directs the sawdust into the system. These hoods can also be clamped on the machine table or mounted in such a way that they can be positioned according to the direction of sawdust and chip deflection. Chip deflection will vary depending upon the direction of shaper cutter rotation or the angle of cut on a radial-arm saw. Of course, flex-hose works very well when repositioning the hood to best capture chips and sawdust.

For bandsaws and stationary disc sanders, a universal dust port can be mounted directly to the lower wheel guard. **See Figure 16**. The angled port directs the flex-hose away from the operator position or tight mounting locations. The universal dust port accepts a 2½" hose and can be increased to 3" with the 2½"x 3" adapter. Note: A small section of 2½" hose and two hose clamps are necessary to connect the universal dust port to the 3" adapter. Two and one half inch hose is a common size for many home shop type vacuums.

In any event, if you plan to add aftermarket hoods to your dust-producing machines, you will most likely need to make some modifications to the hood or to your machine for efficient dust removal.

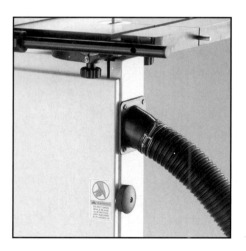

FIGURE 16.
SHOWS A UNIVERSAL DUST PORT MOUNTED TO THE LOWER BANDSAW WHEEL GUARD.

DUCT GROUNDING

If you elect to use plastic pipe or flex-hose, your system must be grounded to safely discharge static electrical build-up. A bare 16 AWG copper grounding wire, which should be stranded and braided for flexibility, must be placed inside the entire duct system, including branch lines. We have found that braided, copper antenna wire which is available in 50 and 100 foot rolls at most electronics stores, is relatively inexpensive and works quite well.

There are a couple of different ways to install the ground wire, particularly when making the connection at branch and main duct locations for a continuously grounded system. One method requires a solder connection at each junction.* However, you must ensure that the soldered lateral wire at the Y or T will not trap chips and sawdust and clog the system. **See Figure 17**.

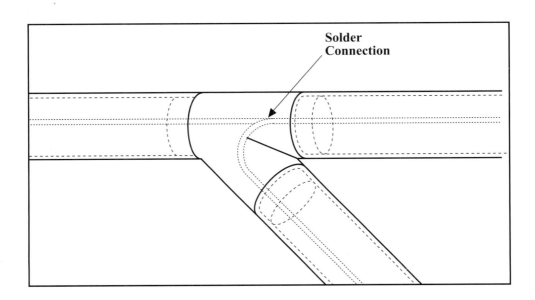

FIGURE 17.
ILLUSTRATES A PROPERLY SOLDERED GROUND-WIRE
CONNECTION INSIDE THE DUCT SYSTEM.
AIRFLOW IS FROM RIGHT TO LEFT.

---

* *Wood Magazine, June 1991*

An alternative method to soldering is to drill small holes in each duct at the junction where two or more wires connect. This may be along a duct or at a fitting where two ducts meet. The wires can be threaded to the outside of the duct system and then simply connected with wire nuts. After threading and connecting the wire ends on the outside, the holes can be sealed with silicone sealant (caulk). This method has the advantage of easily allowing connections on the outside of the system, reducing the chance of clogging and providing a visual means to ensure that the wires stay connected. **See Figures 18 and 19.**

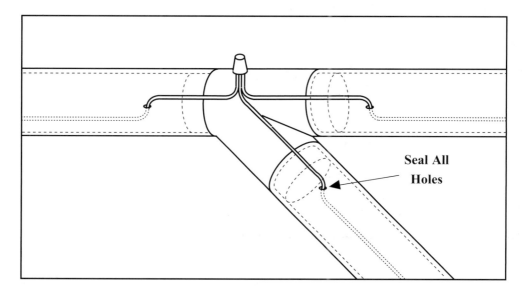

Seal All
Holes

FIGURE 18.

ILLUSTRATES HOW GROUND-WIRES ARE THREADED EXTERNALLY
AND CONNECTED WITH A WIRE NUT.

There have been many articles written on how wire reinforced plastic or rubber hose can solve your grounding worries. This is somewhat misleading since the wire, while present in the hose, is fully insulated. Even if the wire ends are grounded, the wire will not serve to dissipate static electricity build-up inside or outside of the hose; therefore, it does not serve as a proper ground against static electricity. Again, bare wire must be used in order to dissipate static build-up.

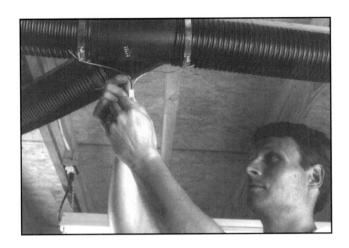

FIGURE 19.
SHOWS CONNECTING GROUND WIRES TOGETHER WITH A WIRE NUT.

In addition to an internal ground wire, we also recommend wrapping a bare wire in spiral fashion around the outside to dissipate any static electricity build-up on the outside of the duct. **See Figure 20.**

FIGURE 20.
SHOWS WRAPPING BARE WIRE AROUND THE OUTSIDE OF THE DUCT.

Connecting branch wires to mainline wires can be achieved simply by cutting the main wire at the branch location and connecting the two ends to the branch wire with a wire nut. **See Figure 21.** The wire must lay flat against the duct with no interruptions between the woodworking machine and dust collector unit. Use electrical tape to secure the wire in place if needed.

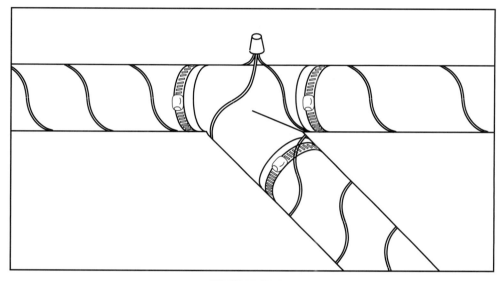

FIGURE 21.

ILLUSTRATES WIRE WRAPPED AROUND THE OUTSIDE OF A Y FITTING.
NOTE WIRE NUT CONNECTION.

Once both ground wires have been installed internally and externally, the wire ends should be connected to each individual woodworking machine by securing them to the machine frame. **See Figure 22.** You should ensure that metal-to-metal contact is made by scraping away paint, if necessary, or by using a frame mounting bolt to double as the connector. The other end of the wire near the dust collector should also be grounded to the dust collector in the same way. You should ensure that each woodworking machine motor, including the dust collector, is continuously grounded to its machine frame and then through the electrical circuit to the grounding terminal in your electric service panel.

To complete your system, caulk any holes that were drilled into the duct with a silicon caulk to ensure lines are air-tight. **See Figure 23.**

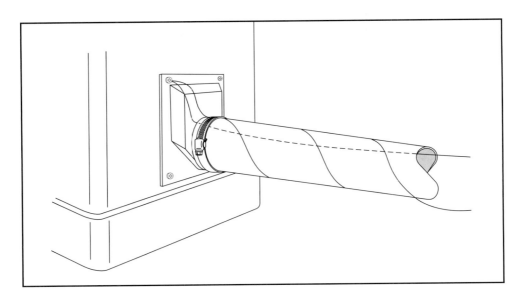

FIGURE 22.
ILLUSTRATES A PROPERLY GROUNDED SYSTEM TYPICAL OF EACH MACHINE LOCATION.

FIGURE 23.
SHOWS SEALING WIRE HOLES. EXTERNAL GROUND WIRES WILL BE TAPED AGAINST THE
DUCT WITH ELECTRICAL TAPE.

Since metal pipe acts as a grounding circuit, static electrical build-up will be dissipated to each woodworking machine frame, assuming that there are no grounding interruptions. Each machine frame should be continuously grounded by a grounding wire from the power cord to the electrical circuit to the grounding terminal in your electric service panel. No grounding wires need to be added except in the case where non-conducting fittings, flex-hose and hoods attach to the metal duct or machine frame. In these instances, all non-conducting components must be bridged or jumped with a ground wire to ensure a completely grounded system. **See Figure 24.**

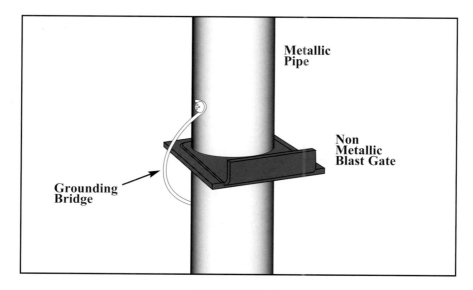

FIGURE 24.

ILLUSTRATES BRIDGING A NON-CONDUCTING FITTING WITH A BRIDGE WIRE.

EQUIPMENT GROUNDING

When connecting your dust collector to a power source, it is very important that it is grounded. The dust collector motor must be wired and grounded according to all federal, state and local electrical codes. If you have questions regarding the proper grounding of your particular dust collector, refer to the owners manual or contact the manufacturer or a licensed electrician.

We want to emphasize that equipment grounding differs from grounding for static electrical charge build-up within a duct system. Equipment grounding protects against the hazard of shock or electrocution caused by a live short in the equipment. Without proper grounding, the electrical charge may pass through a person's body if that person contacts the faulty equipment and completes a ground circuit. Ensure that all equipment is electrically grounded and your setup meets all electrical codes.

⚠WARNING
All electrical equipment with a grounding wire or plug pin must be grounded to protect against shock or electrocution.

SUMMARY

We have attempted to clarify generally accepted design considerations and safety concerns regarding dust collection. We have stressed the importance of planning ahead and identifying as many system variables as possible in order to make informed decisions. Quantifying different system designs will allow you to compare relative efficiencies and weigh personal preferences. A well planned system will adequately handle your dust collection needs, is cost effective and ensures a high degree of safety.

The information presented in this handbook was written to guide you in planning and building a dust collection system. You are solely responsible for your own particular system design and safe construction.

Woodstock International, Inc. assumes no liability regarding the use of its products in the design and construction of any dust collection system or any other system. In no event shall Woodstock International, Inc. be liable for death, injuries to persons, damages to properties, or for incidental, contingent, special or consequential damages arising from the use of its products.

FIGURE 25
SHOWS A COMPLETE BRANCH LINE DUCT IN OPERATION.

ADDITIONAL INFORMATION SOURCES

Industrial Ventilation, A Manual of Recommended Practice 20th Edition by American Conference of Governmental Industrial Hygienists, Edward Brothers, Inc., 2500 South State St., Ann Arbor, MI 48104.

Design of Industrial Exhaust Systems by John L. Alden and John M. Kane, Second Printing, Industrial Press, Inc., 200 Madison Ave., New York, NY 10016.

Fine Woodworking, No. 67, *Clearing the Air, Selecting and sizing a small-shop dust collector* by Roy Berendsohn. Page 70. The Taunton Press, Newtown, CT.

Wood, June 1991, *Central Dust Collection, A simple, affordable system for keeping your shop clean* by Bill Krier, Page 40. Better Homes and Gardens, Des Moines, IA.

Controlling Dust in the Workshop by Rick Peters, First Printing, Sterling Publishing Co., Inc., 387 Park Avenue South, New York, NY 10016.

Woodshop Dust Control by Sandor Nagyszalanczy, First Printing, The Taunton Press, 63 South Main Street, Box 5506, Newtown, CT 06470.

APPENDIX 1

Useful Formulas

Velocity (FPM) = $\dfrac{\text{Volume (CFM)}}{\text{Cross Sectional Area (ft.}^2)}$

Volume (CFM) = Velocity (FPM) x Cross Sectional Area (ft.^{2})

Cross Sectional Area (ft.^{2}) = $\dfrac{\text{Volume (CFM)}}{\text{Velocity (FPM)}}$

Circle Area = 3.142 x Radius^2

Diameter = 2 x Radius

Decimal Equivalents

$\frac{1}{32}$"	.0312	$\frac{17}{32}$"	.5312
$\frac{1}{16}$"	.0625	$\frac{9}{16}$"	.5625
$\frac{3}{32}$"	.0938	$\frac{19}{32}$"	.5938
$\frac{1}{8}$"	.1250	$\frac{5}{8}$"	.6250
$\frac{5}{32}$"	.1562	$\frac{21}{32}$"	.6562
$\frac{3}{16}$"	.1875	$\frac{11}{16}$"	.6875
$\frac{7}{32}$"	.2188	$\frac{23}{32}$"	.7188
$\frac{1}{4}$"	.2500	$\frac{3}{4}$"	.7500
$\frac{9}{32}$"	.2812	$\frac{25}{32}$"	.7812
$\frac{5}{16}$"	.3125	$\frac{13}{16}$"	.8125
$\frac{11}{32}$"	.3438	$\frac{27}{32}$"	.8438
$\frac{3}{8}$"	.3750	$\frac{7}{8}$"	.8750
$\frac{13}{32}$"	.4062	$\frac{29}{32}$"	.9062
$\frac{7}{16}$"	.4375	$\frac{15}{16}$"	.9375
$\frac{15}{32}$"	.4688	$\frac{31}{32}$"	.9688
$\frac{1}{2}$"	.5000	1"	1.0000

APPENDIX 2
Rate Changes of Velocity per Change in Duct Diameter

Duct Dia. (inches)	Required CFM 300 Velocity (FPM)	350 Velocity (FPM)	400 Velocity (FPM)	450 Velocity (FPM)	500 Velocity (FPM)	600 Velocity (FPM)	700 Velocity (FPM)	800 Velocity (FPM)	1000 Velocity (FPM)
2	13751	16043	18335	20626	22918	27502	32086	36669	45837
2.5	8801	10267	11734	13201	14668	17601	20535	23468	29335
3	6112	7130	8149	9167	10186	12223	14260	16297	20372
3.5	4490	5238	5987	6735	7484	8980	10477	11974	14967
3.75	3911	4563	5215	5867	6519	7823	9127	10430	13038
4	3438	4011	4584	5157	5730	6875	8021	9167	11459
4.25	3045	3553	4060	4568	5075	6090	7105	8121	10151
4.5	2716	3169	3622	4074	4527	5432	6338	7243	9054
4.75	2438	2844	3250	3657	4063	4876	5688	6501	8126
5	2200	2567	2934	3300	3667	4400	5134	5867	7334
5.25	1996	2328	2661	2993	3326	3991	4656	5322	6652
5.5	1818	2121	2424	2727	3031	3637	4243	4849	6061
5.75	1664	1941	2218	2495	2773	3327	3882	4436	5545
6	1528	1783	2037	2292	2546	3056	3565	4074	5093
6.25	1408	1643	1877	2112	2347	2816	3286	3755	4694
6.5	1302	1519	1736	1953	2170	2604	3038	3472	4340
6.75	1207	1408	1610	1811	2012	2414	2817	3219	4024
7	1123	1310	1497	1684	1871	2245	2619	2993	3742
7.25	1046	1221	1395	1570	1744	2093	2442	2791	3488
7.5	978	1141	1304	1467	1630	1956	2282	2608	3259
7.75	916	1068	1221	1374	1526	1832	2137	2442	3053
8	859	1003	1146	1289	1432	1719	2005	2292	2865

Minimum velocity for mainlines: 3500 FPM
Minimum velocity for branch lines: 4000 FPM

PLANNING GRIDS

¼ INCH : 1 FOOT

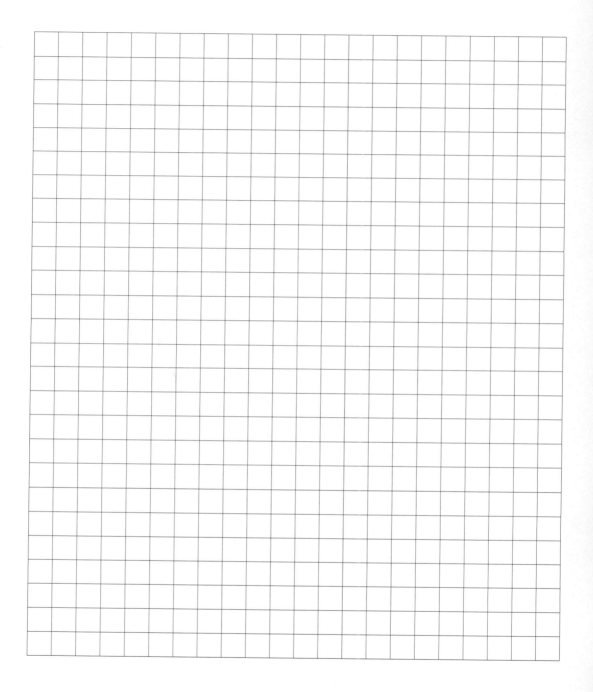

³⁄₈ INCH : 1 FOOT

¹⁄₂ INCH : 1 FOOT

NOTES